WORKBOOK FOR

STATISTICAL METHODS IN EDUCATION AND PSYCHOLOGY

ALBERT K. KURTZ
SAMUEL T. MAYO

Workbook for Statistical Methods in Education and Psychology

Springer Science+Business Media, LLC

A. K. Kurtz
1810 Ivy Lane
Winter Park, FL 32792, USA

S. T. Mayo
School of Education
Loyola University of Chicago
Chicago, IL 60611, USA

Originally published by Springer-Verlag Berlin Heidelberg New York in 1978

9 8 7 6 5 4 3 2 1

ISBN 978-0-387-90324-8 ISBN 978-1-4757-4288-6 (eBook)
DOI 10.1007/978-1-4757-4288-6

CONTENTS

PRETEST OF **MATHEMATICAL ESSENTIALS FOR STATISTICS**

NOTE TO STUDENT: There will be some mathematics in the course you are now starting to take. There will be what some instructors call a little algebra, but some students disagree. How much? That's exactly what this test is designed to let you know. You don't need to hand it in. The authors recommend that you take this test right now. Complete the test. Mark the answers you're sure of and also the ones you aren't sure of. Recheck until you are satisfied. Then (and not until then) turn to the answer key on page 85 and correct your own test. Right below the answer key we will tell you something about your score, but don't read that yet! If this test is to be of any use to you at all, take it, using as much or as little time as you wish and then score it. Good luck!

For each of the items of the test, choose the response you think is correct and PRINT A CAPITAL LETTER that corresponds to it in the space at the left.

ORDER OF ARITHMETIC OPERATIONS

_____ 1. $\frac{1}{8}(32 + 16) + \frac{1}{7}(28 + 14) =$

(A) 38
(B) 26
(C) 24
(D) 12

______ 2. $3\left[\frac{8+3}{3} - 2(3+2)\right] =$

(A) 3

(B) 5

(C) −19

(D) −21

______ 3. $\sqrt{36+64} - (6+8) =$

(A) 16

(B) 12

(C) 0

(D) −4

OPERATIONS WITH FRACTIONS

______ 4. $10\left(\frac{4}{5}\right) + 5\left(\frac{2}{10}\right) =$

(A) 9

(B) $8\frac{1}{5}$

(C) $1\frac{4}{5}$

(D) 1

______ 5. $1 + \frac{k}{n} =$

(A) $\frac{n+k}{n}$

(B) $\frac{1+n}{k}$

(C) $\frac{1+k}{n}$

(D) None of these

______ 6. Which of the following is *not* equivalent to $7 - \frac{8}{3} - \frac{1}{6}$?

(A) $7 - \frac{8 + \frac{1}{2}}{3}$

(B) $\frac{21-8}{3} - \frac{1}{6}$

(C) $\frac{42-1}{6} - \frac{8}{3}$

(D) All are equivalent

______ 7. The reciprocal of the sum of the reciprocals of 5 and 2 is
(A) .10
(B) $\frac{1}{7}$
(C) .70
(D) $\frac{10}{7}$

______ 8. $\frac{48}{.06} =$
(A) 800
(B) 80
(C) 8
(D) .80

______ 9. (3.14) (.002) =
(A) 628
(B) 6.28
(C) .628
(D) .00628

______ 10. Change the proportion .3212 to a percent.
(A) .3212%
(B) 3.212%
(C) 32.12%
(D) 321.2%

______ 11. $\frac{125.75}{7.5} =$
(A) 167.7
(B) 16.77
(C) 1.677
(D) .1677

______ 12. $\frac{.023}{25} =$
(A) .000092
(B) .00092
(C) .0092
(D) .092

______ 13. (.0012) (1,586,247.5) =
(A) 190.3497
(B) 1903.497
(C) 19034.97
(D) 190349.7

______ 14. $\frac{12{,}665.755}{5{,}275.25} =$
(A) .24010
(B) 2.4010
(C) 24.010
(D) 240.10

______ 15. $\sqrt{1125} =$
(A) .3354
(B) 3.354
(C) 33.54
(D) 335.4

______ 16. $\sqrt{112.5} =$
(A) .01061
(B) .1061
(C) 1.061
(D) 10.61

______ 17. $\sqrt{55.40} =$
(A) 74.431
(B) 7.4431
(C) .74431
(D) .074431

______ 18. $\sqrt{5.54} =$
(A) .02354
(B) .2354
(C) 2.354
(D) 23.54

SIGNIFICANT DIGITS

Round off the number 9.35476 to the indicated number of decimal places.

______ 19. To four places
(A) 9.3547
(B) 9.3548

______ 20. To three places
(A) 9.354
(B) 9.355

______ 21. To two places
(A) 9.35
(B) 9.36

______ 22. To one place
(A) 9.3
(B) 9.4

______ 23. To no places
(A) 9
(B) 10

______ 24. $(x - y)^2 =$
(A) $x^2 - y^2$
(B) $x^2 - 2xy - y^2$
(C) $x^2 - 2xy + y^2$
(D) None of these

_______ 25. $(y - b)^2 =$
(A) $y^2 - b^2$
(B) $y^2 - 2yb - b^2$
(C) $y^2 - 2yb + b^2$
(D) None of these

_______ 26. $(n - 1)^2 =$
(A) $n^2 - 1$
(B) $n^2 - 2n + 1$
(C) $n^2 + 2n - 1$
(D) None of these

_______ 27. $(p + q)^3 =$
(A) $p^3 + q^3$
(B) $p^3 + 3(p^2q + pq^2) + q^3$
(C) $(p + q)(p^2 + pq + q^2)$
(D) None of these

OPERATIONS WITH LINEAR EQUATIONS

_______ 28. If $T = 10z + 50$ and $z = -\frac{1}{2}$, then T is equal to
(A) 40
(B) 45
(C) 50
(D) 55

In the next three items (29–31), if $Y = c + dX$, what is Y when

_______ 29. c is 20, d is $\frac{1}{2}$, X is 40?
(A) 30
(B) 40
(C) 50
(D) 60

_______ 30. c is 15, d is $-\frac{1}{5}$, X is 50?
(A) 5
(B) 13
(C) 25
(D) None of these

_______ 31. c is 10, d is $-\frac{1}{3}$, X is 60?
(A) 30
(B) 10
(C) -10
(D) None of these

In the next three items (32–34), if $K + L = U$

_______ 32. what is U when K is 11, L is 9?
(A) 20
(B) 19
(C) 2
(D) 1

_______ 33. what is K when U is 16, L is 9?
(A) 25
(B) 9
(C) 8
(D) 7

_______ 34. what is L when K is 3, U is 3?
(A) 6
(B) 3
(C) 1
(D) 0

In the next two items (35–36), if $Z = \frac{10\,(X - R)}{s} + 50$, *what is* Z *when*

_______ 35. X is 24, R is 96, s is 18?
(A) 90
(B) 10
(C) 0
(D) -10

_______ 36. X is 73, R is 49, s is 12?
(A) 70
(B) 30
(C) 0
(D) None of these

NOTE TO STUDENT: Look back over this test and be sure you answered *every* item. After you have done that, turn to page 85; score your test, count the number of correct answers, and read the **NOTE** below the answers.

CHAPTER 1 THE NATURE OF STATISTICAL METHODS

TRUE-FALSE

_______ 1. Statistical methods make it possible to condense large sets of numbers to manageable size.

_______ 2. Some published figures of the statistical kind can be bewildering and virtually meaningless to the reader.

_______ 3. The only purpose of statistics is to describe figures (or data) which we readily have at hand.

_______ 4. The term *sample* can usually be used interchangeably with the term *population.*

_______ 5. Few educational and psychological measurements are distributed in a manner similar to the normal probability curve.

_______ 6. The normal probability curve is widely used by statisticians.

_______ 7. If you have looked up a value in a table to 4 decimal places you should keep all the places in your calculation.

_______ 8. If rounded to two decimals, each of the following figures will have the same value: 19.491; 19.487632; 19.490005; 19.4943; 19.4850005; and 19.494949.

COMPLETION

NOTE TO STUDENT: The short dashes in the completion items are to indicate that a word or short phrase has been left out. Do not write it there. Instead, write the answer to each item on the long line to the left of it.

________________ 9. In statistics the term we use for a set of figures is ______.

________________ 10. A sample is part of a ______.

________________ 11. The purpose of statistics is to ______ and ______ the interpretation of figures.

____

________________ 12. A kind of statistics used for estimating population characteristics is known as ______.

________________ 13. Differences between two sample means are sometimes tested to determine whether the differences might be attributed to ______.

________________ 14. Ordinarily we would do well to maintain ______ decimal places throughout the steps in a series of calculations.

PROBLEMS

After each number write its correct value when rounded to two decimals.

15. 1.234567 ____________
16. 2.375000...0 ____________
17. 3.68492 ____________
18. 4.67492 ____________
19. 5.885000...0 ____________
20. 6.44448 ____________
21. 7.92501 ____________
22. 8.9250 ____________
23. 9.925 ____________

Round the following to 2 decimal places and put the answer on the line to the right of the number.

24. 12346.49496 ____________
25. 37.49538 ____________
26. .006215 ____________
27. .145000...0 ____________
28. 6.4748495051 ____________
29. .00500...0 ____________
30. 234.500030 ____________
31. 617.49699 ____________

The number, 98.714545001, when rounded to 7; 6; 5; 4; 3; 2; and 1 decimal places is

32. (7) ____________________
33. (6) ____________________
34. (5) ____________________
35. (4) ____________________
36. (3) ____________________
37. (2) ____________________
38. (1) ____________________

The correct values of 5.54544545000...0, rounded to 7, 6, 5, 4, 3, 2, 1, and 0 decimals are

39. (7) ____________________
40. (6) ____________________
41. (5) ____________________
42. (4) ____________________
43. (3) ____________________
44. (2) ____________________
45. (1) ____________________
46. (0) ____________________

The figures in the first column are to be rounded to various numbers of decimal places. Round each of them to 6 decimal places and then record these rounded values in the next column. Then again round the original figures and fill in each of the other columns.

	Figure	6 decimals	5 decimals	4 decimals
47.	2.45454545	________	________	________
48.	3.3344455566	________	________	________
49.	.001325000	________	________	________

Divide and round the quotients to two decimal places.

50. $\frac{1}{8} =$

51. $\frac{3}{8} =$

52. $\frac{5}{9} =$

53. $\frac{5}{11} =$

54. $\frac{2}{7} =$

55. $\frac{5}{7} =$

Perform the indicated operations, keeping all available decimal places.

56. $47.5 + .18 =$

57. $.78\,(69) =$

58. $2\,(.1915) =$

Perform the indicated operations, deferring division as long as possible and always rounding to two decimal places at each step.

59. $10 + \dfrac{-49}{133} =$

60. $112.00 + 5\left(\dfrac{-70}{176}\right) =$

61. $50.50 + \dfrac{-224}{119} =$

62. $48.50 + \dfrac{-205}{119} =$

63. $100\left(\dfrac{17}{70}\right) =$

Divide and round off to two decimal places.

64. $\dfrac{5674}{335} =$

65. $\dfrac{.0247}{659} =$

66. $\dfrac{498{,}217}{969{,}591} =$

Perform the indicated operations.

67. What is 15% of 60? ______________

68. What is 13% of 70? ______________

69. Find the value of U in the equation $U = V + 10$, where $V = 15$.

70. Find the value of Mdn

$$Mdn = l + i\left[\frac{\frac{N}{2} - F}{f_p}\right]$$

where $l = 109.50$
$N = 176$
$F = 81$
$f_p = 27$
$i = 5$

71. Find the value of Y

$$Y = \frac{ab + cd}{a + c}$$

where $a = 4$
$b = 10$
$c = 3$
$d = 5$

72. Find the value of z

$$z = \frac{X - R}{s} \quad \text{where} \quad \begin{array}{l} X = 6 \\ R = 18 \\ s = 4 \end{array}$$

73. Find the value of T

$T = 10z + 50$ where $z = -2.50$

CHAPTER 2 **AVERAGES**

TRUE–FALSE

_______ 1. There is at least one clearcut, legitimate kind of situation in statistical work when we want to group into only two groups.

_______ 2. If you must make a choice, it is better to have too few intervals than too many.

_______ 3. The use of exact limits greatly simplifies the calculation of the mean.

_______ 4. Whether we choose the arbitrary origin near the middle of a distribution or near the lower end does not affect the value of the mean computed from grouped data.

_______ 5. The use of grouped data does not give the same value for the mean as if we had added up all the ungrouped raw scores and divided by the number of scores.

_______ 6. The arithmetic mean is the most valuable of the averages because it is the most stable.

_______ 7. If 17 scores are arranged in order of size, the median will be of the same value as the ninth score from the top.

_______ 8. There are some kinds of data from which it is possible to compute the median but impossible to compute the mean.

_______ 9. The mode is a very stable measure of central tendency.

_______ 10. Numbers are squared in computing the mean.

_______ 11. The median of 4, 6, 7, 8, 1, 11, 7, 11, 13 is greater than the mean of the same scores.

_______ 12. The mean is greater than the median of 5, 10, 7, 8, 6, 6, 12, 7, 5, 13, 9, 6, 5, 6, 6, 10, 5, 7.

_______ 13. The arithmetic mean of 716, 710, 711, 712, 713, 715, 711, 710, 711, and 712 is greater than the median of the same numbers.

COMPLETION

_______ 14. Grouping of scores can be too coarse or too fine. We lose _______ when grouping is too _______ and do not save much labor when grouping is too _______ .

____

____

_______ 15. The _______ of the interval should be set up so as to be a _______ of the interval width.

____

_______ 16. A frequency distribution is a table with basically two columns. The first column shows the _______ while the second column shows the _______ .

____

_______ 17. With an odd number of ungrouped scores in rank order, the middle score will have the same value as the _______ .

_______ 18. When working with whole number scores and i is an odd number, the midpoint of each interval will end with a decimal fraction of _______ , while if i is an even number, the decimal fraction will be _______ .

____

PROBLEMS

19. I add up the scores +3, −2, +7, −12, and −2. Write symbols to show what I have done. _______

20. Compute the arithmetic mean of 7, 5, 8, 7, 10, 9, 7, 12, 7, and 8. _______

21. If we are working with a continuous variable, the arithmetic mean of 293, 297, 294, and 296 will be what value to two decimal places? _______

22. In a list of scores, the lowest two are 311 and 430, while the highest two are 603 and 604. When these are tabulated appropriately, what interval will contain the score 571? _______

23. The shoe sizes of a group of sorority women range from $5\frac{1}{2}$ to 10. What would be the most appropriate interval for tabulating such data? __________

24. The highest and lowest scores in a distribution are 153.19 and 74.65 respectively. What is the size of the most appropriate interval width? What interval would contain the score 123.45? __________ __________

25. If 117 body temperatures ranging from 96.2 to 104.8 are to be tabulated, what will be the exact upper limit of the interval containing the so-called normal temperature of 98.6? __________

26. Assume that you are given data expressed to one decimal place and that you are using an interval of 5. Write the lower score limit, upper score limit, and midpoint of the interval that contains the score which, before rounding, was six and seven-eighths. __________ __________ __________

27. An interval width of twenty degrees is used in tabulating temperature data, of which the individual readings are given to one-tenth of a degree. What is the midpoint of the interval in which the temperature 234.5 would be tabulated? __________

28. If one interval in a distribution is 15.00-19.99 and if the lowest score is negative, what is the midpoint of the interval in which the score 38.17 is tabulated? __________

29. Compute the median of 7, 5, 8, 7, 10, 9, 7, 12, 7, and 8. __________

30. Compute the arithmetic mean and median of the scores 3, 6, 12, 24, 48, 96, 192, 384, 768, and 1536. $\bar{X}$ __________ *Mdn* __________

31. Compute the arithmetic mean and median of the scores 4, 8, 16, 32, 64, 128, 256, and 512. $\bar{X}$ __________ *Mdn* __________

32. The mode of 7, 5, 8, 7, 10, 9, 7, 12, 7, and 8 is what value? __________

33. On a used car lot, the average price of 3 Cadillacs is \$5900; that of 6 Buicks is \$5100; that of 9 Pontiacs is \$4300; and that of 12 Chevrolets is \$3700. What is the average price of all these cars? __________

34. The mean test score of 3000 males is 37.08 and of 7000 females is 33.08. What is the mean for the 10,000 people? __________

35. Using the data in Group I of Data Set A*, calculate the following statistics using ungrouped scores.

Statistic	Math Pretest	Math Opinion	Final Exam
a. Median	__________	__________	__________
b. Mean	__________	__________	__________

36. Using the data in Data Sets C* and D*, calculate the following statistics using x' score methods for the mean and median.

Statistic	Data Set C	Data Set D
a. Mean	__________	__________
b. Median	__________	__________
c. Mode	__________	__________

37. Using Data Set C and Data Set D, plot the following on graph paper with 5 small squares to the inch, using the rule given on page 34 of the text that the graph should be around twice as wide as it is high.

   a. Histogram (plot with real limits for sides of rectangles but use midpoints to label intervals).

   b. Frequency polygon.

   c. Cumulative percentage curve.

*Data Sets A, C, D, and the others are given on pages 75–84 near the end of this *Workbook*.

CHAPTER 3 THE STANDARD DEVIATION

TRUE-FALSE

_______ 1. If the means of two distributions are identical, then the variabilities must also be identical.

_______ 2. The standard deviation is a measure of central tendency.

_______ 3. The standard deviation is usually interpreted in connection with the median.

_______ 4. The formula $s = \sqrt{\frac{\Sigma x^2}{N - 1}}$ is useful for computing the standard deviation.

_______ 5. The variance is always the square of the standard deviation.

_______ 6. The variance can never be negative.

_______ 7. The more homogeneous a distribution is, the smaller the standard deviation.

_______ 8. $\Sigma fx'^2 = \Sigma f(x')^2$.

_______ 9. The Charlier check is a good method for finding out whether the last figure (and all others) in a square root is accurate.

_______ 10. Standard scores may be used as a statistical device to combine two different scores into a composite score.

_______ 11. Numbers are squared in computing the standard deviation but not in computing the average deviation.

______ 12. The square root of 123,456, correct to two decimals, is 111.11.

______ 13. A T score of 66 is above average.

______ 14. A T score of 68 is higher than a standard score of 2.

COMPLETION

____________________ 15. In order to describe a set of scores in summary fashion, the two best items of information are usually the ______ and the ______ .
____

____________________ 16. The best measure of variability is the ______ .

____________________ 17. The standard deviation is very close to the square root of the average of the ______ .

____________________ 18. In the standard deviation formula the symbol x means ______ .

____________________ 19. The sum of the deviations of any set of scores from their mean is always equal to ______ .

____________________ 20. The symbol σ_N in this textbook refers to the standard deviation of ______ .

____________________ 21. In order to combine two or more kinds of raw scores into composite scores, we need to equate the ______ of the different kinds of scores. We can do this by the use of ______ .
____

____________________ 22. Standard scores, designated by z, always have a mean of ______ and a standard deviation of ______ .
____

PROBLEMS

23. If $N = 20$; $\Sigma X = 492$; and $\Sigma X^2 = 12{,}432$; compute the mean and standard deviation. $\overline{X}$ ______ s ______

24. Compute the mean and the squared standard deviation of the following time scores (in seconds): 63, 52, 65, 50, 68, 56, 59, 64, 58, and 65. $\overline{X}$ ______ s^2 ______

25. What is the standard deviation of the following ten scores: 5, 3, 1, 2, 0, 1, 3, 4, 1, 0? ______

26. Find the standard deviation of 7, 5, 8, 7, 10, 9, 7, 12, 7, and 8. ______

27. The standard deviation of the scores 5, 4, 3, 3, 3, 2, 2, 2, 1, 0 is the square root of what number? ______

28. What is the square of the standard deviation of 4, 7, 9, 0, 7, 6, 5, 2, 8, and 6? ______

29. What is the square of the standard deviation of 8, 14, 18, 0, 14, 12, 10, 4, 16, and 12? ______

30. Without grouping the data, compute the standard deviation of the following personality inventory scores: 120, 122, 105, 115, 121, 107, 122, and 112. ______

31. Without grouping the data compute the standard deviation of 10, 13, 17, 12, 18, 14, 2, 7, 13, and 4. ______

32. The following scores are each 10 units smaller than those in the preceding problem. What is their standard deviation? The new scores are 0, 3, 7, 2, 8, 4, -8, -3, 3, and -6. ______

33. Using the data in Data Set A, calculate the standard deviation from the ungrouped data in Group I.

	Group I		
	Math Pretest	*Math Opinion*	*Final Exam*
Standard Deviation	______	______	______

34. Using the data in Data Sets C and D, calculate the standard deviation by formula (3.3) for grouped data.

	Data Set C	Data Set D
Standard Deviation	__________	__________

35. Using the methods of combining two different raw scores for each person by transforming to standard scores (z scores) and averaging as described in Chapter 3, average the Arithmetic and Mathematics scores for the 22 persons in Class #1 of Data Set B.

36. A school psychologist in Bird-in-Hand, Pennsylvania, has a student with a reading T score of 38. An equally good reader in the nearby town of Paradise has what z score?

37. Record the square roots, correct to two decimal places, at the right of each of the following numbers.

1.702874395	__________	19,752.90	__________
1776	__________	32.32	__________
19,753,002	__________	1,652.02	__________

38. The following square roots are alleged to be correct. If they are correct to the number of places given, indicate this with a check mark (√). If they are incorrect, indicate whether the true square root is larger or smaller than the figure given.

    a. Square root of .67448975 is .8215 __________

    b. Square root of 67448.975 is 821.3 __________

    c. Square root of 1470.0739 is 12.12 __________

    d. Square root of .1468888 is .3833 __________

39. If $\overline{X} = 32$; $s = 16$; and $N = 10$, what is the value of the variance? __________

40. What is the variance of the numbers 13, 10, 11, 8, 12, 11, 14, and 9? __________

41. Someone looks at the scores +3, -2, +7, -12, and -2, ignores their signs, and adds the numbers. Write symbols to show what he has done. __________

42. Choose any formula which is appropriate and correctly use it to obtain values for the mean, median, and the standard deviation for Data Set H.

$\overline{X}$ __________

Mdn __________

s __________

CHAPTER 4 NORMAL PROBABILITY CURVE

TRUE–FALSE

_______ 1. The use of the normal probability curve is almost completely confined to psychological and educational data.

_______ 2. Distributions of IQs for a heterogeneous population rarely approach a normal curve shape.

_______ 3. An ordinate of a normal probability curve is its deviation divided by the standard deviation.

_______ 4. As defined by the mathematician, the normal probability curve never reaches the baseline at either the extreme low or extreme high ends.

_______ 5. Pascal's triangle can sometimes be used to obtain coefficients of the binomial expansion.

_______ 6. The normal probability curve can be approximated by coin tossing.

_______ 7. A particular normal probability curve can be completely specified by its N, mean, and standard deviation.

PROBLEMS

8. Which of these events is more likely: getting 4 tails when 5 coins are tossed or getting 1 head when 6 coins are tossed? If you can do so, use Pascal's triangle to show the basis for your answer. If you can't, use any method to show it. ______

9. A retired congressman tosses 6 unbiased coins several thousand times. In what proportion of the tosses will he get 2 or fewer tails? ______

10. In how many different ways can exactly 3 heads be obtained when a penny, nickel, dime, quarter, half dollar, and dollar are tossed? ______

11. A multiple choice test contains 20 5-choice items. In how many different ways can a person get exactly 18 of the items right? ______

12. In a normal distribution, what percentage of the cases fall between 24 and 28 if $\bar{X} = 24$ and $s = 2$? ______

13. Given a normal distribution with a mean of 102.28 and a σ of 26.25, what limits will include the middle 49% of the cases? ______

14. In a normal distribution, how many standard deviations from the mean are the limits that include the middle 95% of the cases? ______

15. What proportion of the cases in a normal distribution are included between the ordinates going through the two points of inflection? ______

16. Taking the area under the normal probability curve as 100%, find the percentage of the total area correct to 2 decimal places between ordinates at the following locations:

a. -1.23σ to $+2.45\sigma$ ______

b. $-.72\sigma$ to $-.62\sigma$ ______

c. -1.11σ to $-.99\sigma$ ______

d. zero to $+1.00\sigma$ ______

e. zero to $+1.75\sigma$ ______

f. $-.25\sigma$ to $+.75\sigma$ ______

g. $+.50\sigma$ to $+1.50\sigma$ ______

h. -1.75σ to $-.75\sigma$ ______

i. mean to $+2.34\sigma$ ______

j. mean to -1.23σ ______

k. -2.96σ to $+.06\sigma$ ______

l. -1.56σ to + infinity ______

m. $-.84\sigma$ to − infinity ______

17. In a normal distribution with $\overline{X} = 240$ and $s = 20$, what percentage of the cases have scores between 210 and 290? __________

18. If height is normally distributed, how many boys in a class of 200 will be more than 2.34*s* below the mean? __________

19. The Registrar of a state university reports that there are 10,832 students on campus during a certain semester. If reading speed is normally distributed, how many of them are between 1.63 standard deviations below their mean and one standard deviation above it? __________

20. If IQs have a mean of 100 and a standard deviation of 17 in California, how many of 2,000,000 California children will have IQs of 140 or over? (Hint: An IQ of anything between 83.50 and 84.50 is called 84.) __________

21. We have a normally distributed population mean of 50 and a standard deviation of 15. What proportion of the scores are below 80? __________

22. Assume that male basketball players have a mean height of 74 inches with a standard deviation of 3 inches and that local girls have a mean height of 66 inches with a standard deviation of 2 inches. What proportion of basketball players would have to wear special shoes in order to be taller than she is if they all dated a girl who is 2.5 standard deviations above the girls' mean? __________

23. What percentage of students make scores above -1.0σ? __________

24. How many people out of 10,000 should have scores below $+1.60\sigma$? __________

25. If the mean IQ of children in Shickshinny is 99.50 and the *s* of the IQs is 14, what percentage of the Shickshinny children will have IQs of 107 to 120 inclusive? (See hint in Problem 20.) __________

26. If the mean IQ of children in Lick Skillet is 111 and the *s* of the IQs is 10, what percentage of the Lick Skillet children will have IQs of 110 to 125 inclusive? __________

27. If the IQs of Chopsoppy children are normally distributed with a mean of 101 and a standard deviation of 15, at what IQs will there be half as many children as there are at 92? __________

28. What proportion of the normal probability curve lies in the tail beyond the point where the curve is only $\frac{1}{10}$ as high as it was at the mean? __________

CHAPTER 5 STATISTICAL INFERENCE

TRUE–FALSE

_______ 1. The sampling distribution of means can be normal only if the population is normal.

_______ 2. Traits which can be measured are called variables; those which cannot be measured are called parameters.

_______ 3. The arithmetic mean of the times taken by 118 men in running the 100 yard dash is a statistic rather than a parameter.

_______ 4. Sample statistics are often used as estimates of population parameters.

_______ 5. Other things being equal, the greater the number of cases, the greater will be the standard error of the mean.

_______ 6. As the sample size increases, *df* decreases.

_______ 7. When N increases, the standard error of the mean also increases but not necessarily at the same rate.

_______ 8. When sample size becomes very large, the theoretical t distribution approaches the normal probability distribution rapidly.

_______ 9. When $t = .24$, the null hypothesis will probably be rejected.

_______ 10. For an N of 17, the t distribution gives a more accurate interpretation of the standard error of the mean than the normal probability curve does.

______ 11. A difference such that it or a greater difference could have happened by chance only once in 20 trials is said to be significant at the .05 level.

______ 12. Higher chances of being correct are needed in gambling than in scientific decisions.

______ 13. The standard error of the median is greater than that of the mean.

______ 14. The standard error of the standard deviation is always less than the standard error of the mean of the same distribution.

______ 15. The purpose of testing any hypothesis is to determine if we can prove it.

______ 16. Null hypotheses are usually more difficult to test than other kinds of hypotheses.

______ 17. The basic assumption involved in the null hypothesis is that chance fluctuations might explain whatever differences are found.

______ 18. The significance level gives directly the proportion of Type I errors.

COMPLETION

____________ 19. When the mean, median and standard deviation are computed from samples, these results are called ______ .

____________ 20. When the mean, median, and standard deviation are computed from an entire population, these results are called ______ .

____________ 21. A statistic is usually symbolized with a (Roman/Greek) ______ letter while a parameter is usually symbolized
____________ with a (Roman/Greek) ______ letter.

____________ 22. Parameters are associated with a ______ , while statistics are associated with a ______ .

____

____________ 23. A sample drawn so that every individual in the population has an equal and independent chance of being included is called a ______ sample.

____________ 24. If many samples are drawn from the same population and some statistics (such as $\overline{X}$, *Mdn*, or *s*) are computed for each sample, then the distribution of the values of one of these statistics would be called a ______ distribution.

____________ 25. The tilde sign (symbolized as ˜) over a symbol indicates that the value has been ______ .

__________ 26. If the sample standard deviation is divided by the square root of the number of cases, we obtain what is called the ______ .

__________ 27. When N is large, the t distribution is similar to a ______ .

__________ 28. When N is small, the t distribution has a (higher/lower) ______ frequency 2 or 3 standard deviations from the mean than does the normal distribution.

__________ 29. When the sample size is small, the df is (small/large) ______ .

__________ 30. When the df is small, the values of t required for significance are (small/large) ______ .

__________ 31. When df is infinite, the t distribution is ______ .

__________ 32. In a test of significance using t, the results for a particular df are significant at the .05 level if the obtained t is (smaller/larger) ______ than the tabled t in the .05 column.

__________ 33. If we establish some significance level in an experiment (such as .05, .01, or .001) and then find our t value to be smaller than the tabled value, we would (retain/reject) ______ our null hypothesis.

__________ 34. When we set up confidence limits around an obtained sample mean, we make what is known as an ______ estimate.

__________ 35. When we establish confidence limits for a parameter, we engage in statistical ______ .

__________ 36. To estimate a mean and median with equal accuracy from a sample would require (less/more) ______ cases for the median than for the mean.

__________ 37. In a Type I error our null hypothesis is actually (true/false) ______ but we decide to (retain/reject)
__________ ______ it.

__________ 38. In a Type II error our null hypothesis is actually (true/false) ______ but we decide to (retain/reject)
__________ ______ it.

PROBLEMS

39. If $s = 7.2$ and $N = 145$, what is the standard error of the mean? ______

40. A standard error of a mean based on 99 cases is .09. The experimenter wishes the standard error to be .01. How many cases does he need? ______

41. In an experiment the standard error of the mean was found to be .000008 for a group of 17 cases. Although the standard error looks small to some people, the experimenter would like to get his standard error down to about .000002. What sound advice can you give him with regard to his objective? ______

42. A class of 37 students has a mean IQ of 108 and a standard deviation of 18. Are these students significantly superior to the general population mean of 100?

    a. At the .01 level? ______

    b. At the .05 level? ______

43. Establish .95 confidence limits for a mean of 60 with a standard deviation of 5 and a sample size = 41. ______

44. If $\overline{X} = 68.02$; $Mdn = 67.89$; $s = 13.68$; and there are 60 cases, what is the standard error of the median? ______

45. If $s = 7.16$, $N = 72$, and $s_{\overline{X}} = .84$, what is the value of the standard error of the standard deviation? ______

46. Two students are using data selected at random from the same population: G computes a mean from his 170 cases. H likes medians better than means. How many cases will H need to select in order to compute a median that will be as reliable as G's mean? ______

47. Scientists sometimes establish at the .001 level that the null hypothesis is false. Under what circumstances could a scientist similarly establish that it is true? ______

CHAPTER 6 PERCENTILES AND PERCENTILE RANKS

TRUE–FALSE

_______ 1. In a class of 25, the tenth percentile is the score made by the third lowest person.

_______ 2. The standard error of a percentile is constant for different percentiles in a given distribution.

_______ 3. In a grouped frequency distribution percentiles are computed from cumulative frequencies.

_______ 4. The average of the tenth and ninetieth percentiles is known as *D*.

_______ 5. *D* is usually much larger than *s*.

COMPLETION

_______________ 6. A percentile is a (point/distance) _______ .

_______________ 7. The number of percentiles is _______ .

_______________ 8. Deciles, quartiles, and the median are all _______ .

_______________ 9. A percentile rank is expressed as a _______ .

_______________ 10. The median is equivalent to the _______ percentile.

_______________ 11. Write the symbol for a percentile which is equal to the third or upper quartile. _______

__________ 12. P_{90} and P_{10} are points, while D is a ______.

__________ 13. Q_1 and Q_3 are points, while Q is a ______.

__________ 14. The least useful measure of dispersion is the ______.

PROBLEMS

15. How many people out of 10,000 should have percentile ranks below 57? __________

16. What is the percentile rank of a student having a z score of -1.34? __________

17. What is the percentile rank of a student having a z score of .25? __________

18. Compute Q (the quartile deviation) for the following scores: 8, 9, 12, 6, 5, 11, 7, 9, 8, 10, 9, 7, 8, 9, 6, 8, 5, 10, 7, and 11. __________

19. Compute Q (the quartile deviation) for the following scores: 20.8, 20.6, 20.6, 20.8, 21.2, 19.6, 29.8, 20.8, 20.2, 21.2, 20.0, 19.4, 22.4, 21.2, 21.4, 19.6, and 21.2. __________

20. Compute D (the 10 to 90 percentile range) for the following scores: 106, 98, 107, 106, 112, 97, 100, 106, 106, 101, 104, 109, 105, 99, 98, 106, 104, 103, 103, and 104. __________

21. Using the data in Data Sets C & D, calculate the following statistics from the grouped data.

Statistic	Data Set C	Data Set D
a. D	__________	__________
b. Q	__________	__________

22. Given a normal distribution with $\bar{X} = 97.53$ and $s = 8.00$. Solve for P_{10}, P_{90}, and D (the 10 to 90 percentile range).

23. Compute the percentile ranks for the midpoints of the intervals in the following data sets.

 a. Data Set C.

 b. Data Set D.

 c. Data Set E.

24. What is the proportion of a normal distribution included between the 60th percentile and a point 1.54 standard deviations above the mean? ____________

25. What is the proportion of a normal distribution above the lower quartile but below the point one-half standard deviation below the mean? ____________

26. At what point with a positive standard score is the normal probability curve only one-third as high as it is at P_{39}? ____________

CHAPTER 7 SKEWNESS AND TRANSFORMED SCORES

TRUE-FALSE

______ 1. When a test of significance shows that the skewness is not significantly different from zero, we may rightly conclude that the skewness is zero.

______ 2. Kurtosis, as well as skewness, concerns departures from symmetry.

______ 3. A distribution with large kurtosis may be either more peaked or less peaked than a normal distribution.

______ 4. In a leptokurtic curve, there are more scores more than $3s$ away from $\bar{X}$ than there are in a normal distribution with the same $\bar{X}$ and s.

______ 5. There are only a very few ways in which one set of scores can be transformed into another kind of scores.

______ 6. Sometimes skewness can be reduced by some kind of transformation of the raw scores.

______ 7. Sometimes transformed scores can show higher correlations with other variables than the original scores did.

______ 8. If the midpoint of the range covered by D coincides with the point above which 50% of the cases lie, the distribution is probably skewed.

COMPLETION

______________ 9. The kind of kurtosis in a normal distribution is called ______.

______________ 10. When dealing with weights or volumes, the most likely transformation is ______.

______________ 11. When the median of a sample of scores is lower than the mean, the distribution is said to be ______ skewed.

______________ 12. Two characteristics of z scores which may prove objectionable are that they have ______ and ______.

____

______________ 13. The T score distribution is obtained by (a linear/an area) ______ transformation.

PROBLEMS

14. a. Test Data Set E for skewness and kurtosis using Formulas (7.1), (7.2), (7.3), and (7.4).

b. Interpret your results from part a.

15. a. Compute the skewness measure Sk by Formula (7.1) for both distributions (Raw Scores and T scores) from Supplement 7, which are reproduced here.

Raw Scores	f	T Scores
300–309	1	77
290–299		74
280–289		74
270–279	1	73
260–269	4	69
250–259	4	66
240–249	8	63
230–239	5	61
220–229	10	58
210–219	6	57
200–209	8	55
190–199	2	54
180–189	2	54
170–179	6	53
160–169	4	52
150–159	7	51
140–149	14	49
130–139	12	47
120–129	8	45
110–119	11	43
100–109	8	40
90–99	12	37
80–89	3	32
70–79	2	30
60–69	1	27
50–59	1	23

b. Test the significance of the two Sk's by use of s_{Sk} in Formula (7.2).

c. Compute the kurtosis measure Ku by Formula (7.3) for both distributions.

d. Test the significance of the two Ku's by use of s_{Ku} in Formula (7.4).

e. Interpret the results of parts a, b, c, and d.

CHAPTER 8 PEARSON PRODUCT MOMENT COEFFICIENT OF CORRELATION

TRUE–FALSE

_______ 1. If we have a clear choice, we would choose Pearson product moment correlation over any of the other methods of correlating two variables.

_______ 2. A negative correlation indicates that large amounts of one variable tend to accompany small amounts of the other variable.

_______ 3. The Pearson product moment coefficient of correlation measures the extent of relationship, but it does not show which variable is the cause of the other one.

_______ 4. In using a correlation chart, the data cannot be grouped by the use of intervals.

_______ 5. The larger one of the arbitrary origins is, the larger will be the correlation coefficient.

_______ 6. In computing correlations from grouped data, raising both arbitrary origins lowers the correlation.

_______ 7. The z score formula for Pearson r is a good computing formula.

_______ 8. To compute Σxy, we multiply Σx by Σy.

_______ 9. The two coefficients r_{xy} and r_{yx} would have the same computed value for variables X and Y.

_______ 10. Sometimes a *dependent* variable is called the *criterion* variable.

______ 11. A person who had computed r on data with non-linear regression might rightfully be accused of fraud.

______ 12. In computing r we must assume normality.

______ 13. When we can say that an obtained value of r is not significant at the .05 level, this means we have proved that there is no correlation.

______ 14. A reliability coefficient of .60 is high, but a validity coefficient of .60 is low.

______ 15. Pearson product moment correlation coefficients cannot be computed without plotting the data on some kind of correlation chart.

______ 16. For a given value of r, the greater N is, the smaller the standard error of r will be.

______ 17. An r of .37 based on 37 cases is significant at the 5% level but not at the 1% level.

______ 18. If $r = .80$ and $N = 101$, then $s_r = .06$.

______ 19. A correlation of .50 for 40 cases indicates some degree of positive association rather than a pure chance relationship.

______ 20. The reliability of a test shows how consistent the scores on it are.

______ 21. If a test is not valid, it cannot possibly be reliable.

______ 22. The validity of a test is defined as the extent to which it predicts what it is supposed to predict.

______ 23. A correlation coefficient between a test and a criterion that it is intended to predict is called a validity coefficient.

______ 24. Usually, the best way to determine the reliability of a test is to give it, wait a suitable time, and give an alternate form of it.

______ 25. The correlation between two halves of a test will usually be lower than the correlation between two forms of the test.

______ 26. An estimate of the amount by which a test should be lengthened can be obtained by using the Spearman–Brown formula.

______ 27. The upper limit of the correlation of an ideal test with a test having a reliability of r_{1I} is $\sqrt{r_{1I}}$.

______ 28. When a correlation between the same variables is computed for a heterogeneous and for a homogeneous group, the former correlation will usually be larger than the latter.

COMPLETION

_______________ 29. Sometimes when one has all the correlations among a set of variables, one can explain the relationships in terms of a smaller number of quantities, which are called ______ .

_______________ 30. The process by which one obtains the smaller number of quantities to explain the correlations among a set of variables is called ______ .

_______________ 31. For plotting a scatter diagram, we need for each tally mark ______ scores.

_______________ 32. The intersection of any particular row with any particular column in a scatter diagram is called a ______ .

_______________ 33. In Chapter 8 of the text the formula for correlation which *looks* the *simplest* is one which uses ______ scores.

_______________ 34. In a scatter diagram in which the variables are grouped into intervals, if the means of the columns do not deviate systematically from a straight line, we say the regression is ______ .

_______________ 35. If we compute r on data with non-linear regression, the obtained value of r is an ______ of the actual relationship.

_______________ 36. If we slice a normal correlation surface parallel to the X axis or Y axis, the shape of the cut will be ______ .

_______________ 37. If we slice through the normal correlation surface parallel to the horizontal plane, the shape of the cut is ______ .

_______________ 38. If there is no consistent relation at all between pitch discrimination and numerical ability, the correlation between these two variables will be ______ .

MATCHING

Directions: Indicate whether Reliability *or* Validity *applies to the situation at left by printing an* "R" *or a* "V" *in the blank.*

	Situation	Test Characteristic
______	39. Same test given same people on two different occasions.	(R) Reliability
______	40. Test correlated with a non-test measure.	(V) Validity
______	41. Alternate forms of same test given same people on two occasions.	
______	42. Test given once, split into halves and scores correlated.	
______	43. Kuder–Richardson formulas.	

PROBLEMS

44. Given $\Sigma X = 810$; $\Sigma X^2 = 5620$; $\Sigma Y = 1060$; $\Sigma Y^2 = 8800$; $\Sigma XY = 6800$; and $N = 130$, compute r. ______

45. Compute r from the following data: $N = 212$; $\Sigma x = 0$; $\Sigma x' = 702$; $\Sigma X = 1126$; $\Sigma X^2 = 7314$; $\Sigma Y = 946$; $\Sigma Y^2 = 5270$; and $\Sigma XY = 5950$. ______

46. A group of Puritan children participated in a spelling bee, which Abner won with Daisy right behind him. The number of words which each child spelled correctly is given below. The number of absences from church in 1642 is also given. Many absences were unavoidable, due to being snowed in, captured by Indians, etc. What is the relation between failure to attend church and spelling ability?

Christian Name	Spelling	Absences
Abner	16	2
Benedict	2	0
Charity	10	3
Daisy	15	0
Enoch	0	5
Faith	10	10
Gabriel	0	4
Hope	4	5
Isabel	1	0
Jesse	4	13
Katrinka	8	1
Luella	3	3

______

47. Select an appropriate formula for computing the Pearson product moment coefficient of correlation between the following test scores.

English	15	13	11	9	17	7	15	14	16	11
Spanish	9	10	5	7	9	6	11	8	9	4

____

48. Before computing a correlation, I subtract 30 from every X score, thereby reducing the mean from 87.14 to 57.14. I also double all Y scores (which were shoe sizes) to eliminate fractions. What effect will these operations probably have on the Pearson r? ______________

49. A kilogram is 2.2 pounds and a centimeter is .4 inch. The correlation between the height in centimeters and weight in kilograms of statues of Greek goddesses is .8660. What is your best estimate of what the correlation would be if the measurements were to be made in inches and pounds? ______________

50. Using 42 cases, an investigator obtains these figures: $\Sigma x' = 450$; $(\Sigma x')^2 = 202500$; $\Sigma x'^2 = 5376$; $\Sigma y' = 304$; $(\Sigma y')^2 = 92416$; $\Sigma y'^2 = 2514$; $\Sigma x'y' = 3506$; $\overline{X} = 47.14$; $\overline{Y} = 50.71$.
r is equal to a numerator divided by the square root of something. In this case what is the numerator? The denominator is the square root of what number? ______________ ______________

51. Is the correlation of .7826 based on $N = 212$ significantly different from zero? ______________

52. When we test the significance of a correlation coefficient and find that the null hypothesis is not rejected, how should we interpret our findings? ______________

53. If $N = \overline{X} = \overline{Y} = 38$; $s_x = s_y = 3.8$; and $r = .38$, show how to use t to determine whether this correlation is inconsistent with the null hypothesis that the population r is zero. ______________

54. With 26 employees of government alphabetical agencies, a correlation of .40 is obtained between salary and a test designed to measure ingenuity in thinking up possible functions for proposed hypothetical agencies identified solely by their imaginary alphabetical initials. Analyze these results and justify the conclusions you draw. ______________

55. a. Compute one Pearson r from one pair of the variables given in Group I of Data Set A. Use the variables specified by your instructor. (If he doesn't specify, then compute r_{13} from Mathematics Pretest and Final Examination. For convenience, let Math Pretest be Variable No. 1; Math Opinion be Variable No. 2; and Final Exam be Variable No. 3.)

b. Test your coefficient for significance.

56. a. Compute two Pearson r's from Data Set B for the three classes combined. Use the variables specified by your instructor. If he doesn't specify, then compute r_{15} (between Aptitude and First Year Average) and r_{45} (between English and First Year Average). (For convenience, let Aptitude be Variable No. 1; Arithmetic be Variable No. 2; Mathematics be Variable No. 3; English be Variable No. 4; and First Year Average be Variable No. 5.)

b. Test each coefficient for significance.

CHAPTER 9 REGRESSION EQUATIONS

TRUE–FALSE

______ 1. If both regression lines are linear, but one of the distributions is skewed while the other is not, one variable can probably be predicted more accurately than the other one can.

______ 2. If the standard error of estimate of Y (in inches) is less than the standard error of estimate of X (in pounds), we can predict height from weight more accurately than we can predict weight from height.

______ 3. In using regression equations to relate two variables, we often wish to predict in both directions.

______ 4. If height and weight correlate +.60 and if a man is 90% as tall as the average, his predicted standard score on weight will be positive.

______ 5. If a senior has a standard score of +2.00, his predicted standard score on any other variable can never be greater than +2.00.

______ 6. If r between height and weight is -1.00, and if Quentin is two inches above the average height, he will probably be two pounds above the average weight.

______ 7. If a woman is sufficiently far above the mean on one variable she will probably be above the mean on the second variable regardless of the correlation between them.

______ 8. If Y is more variable than X, the regression coefficient used to predict Y will be larger than the regression coefficient used to predict X.

______ 9. From the data needed to write a linear regression equation to predict X from Y, it is always possible to write a regression equation to predict Y from X.

______ 10. The standard error of estimate of Y scores will be zero either if s_y is zero or if r is +1.00 or if r is −1.00.

______ 11. The standard error of estimate provides a useful standard for determining when errors in computation or in rounding off decimals become excessive.

______ 12. If r is negative, the standard error of estimate will always be less than the corresponding standard deviation.

______ 13. If the correlation does not equal zero, the standard error of estimate will always be smaller than the standard deviation.

______ 14. There are two standard errors of estimate for a set of data giving scores on two variables.

______ 15. The standard error of estimate is a standard deviation.

______ 16. If the standard error of estimate of Y scores is .68, then 68% of the predicted Y scores may be expected to differ from their actual scores by .68 or less.

______ 17. If regression is linear, a straight line can be drawn which will come close to the means of the rows in the correlation table.

______ 18. The standard error of estimate furnishes a valid estimate of errors made by the computer.

______ 19. All regression lines pass through the point whose coordinates are the means of the two variables.

______ 20. If the correlation between education and earnings is zero, the earnings predicted from a regression equation will be exactly the same for a group of college graduates as for a group who stopped at the fifth grade.

______ 21. If one correlation coefficient is +.49 and another is −.49, their coefficients of alienation will also have the same absolute values with opposite signs.

______ 22. As the validity increases, the coefficient of forecasting efficiency also increases.

______ 23. E is always larger than the corresponding k.

______ 24. If we took all the positive correlations in this book and correlated them with the corresponding coefficients of forecasting efficiency, this new correlation would be positive.

______ 25. If $r = .50$, both k and E will also equal .50.

______ 26. A test with a validity of only .40 will be quite useful if the selection ratio is very low.

______ 27. If people with the highest test scores are hired, and if the test has a high validity, the smaller the proportion hired, the greater will be the proportion of new employees who are considered satisfactory.

______ 28. The coefficient of correlation is usually larger in absolute value than the coefficient of determination.

COMPLETION

____________________ 29. A ______ equation is an equation for predicting the ______ value of one variable from the ______ value of another variable.

____

____

____________________ 30. The simplest and most fundamental kind of regression equation is one based upon ______ scores.

____________________ 31. The symbol b_{yx} is known as a ______. It indicates that variable ______ is predicted from variable ______.

____

____

____________________ 32. The symbol r_{yx} functions as a ______ in a regression equation used to predict z scores on variable y from z scores on variable x.

____________________ 33. If we know the values of the two regression coefficients we can find the numerical value of the correlation coefficient by taking the square root of the ______.

____________________ 34. Equal variability in all the arrays of a scatter diagram is called ______.

____________________ 35. The standard error of estimate is the standard deviation of the ______.

____________________ 36. We can sometimes obtain linear regression by a process of ______ one or more of the variables.

____________________ 37. We can plot a regression line with ______ points but it is more desirable to have ______ points as a check.

____

____________________ 38. When $r = \pm 1.00$, the angle between the two regression lines is ______ degrees.

__________ 39. The one sense in which r can be interpreted as a proportion is in a situation in which it tells us what proportion the predicted standard score $\tilde{z}_y$ will be of the ______.

__________ 40. The smaller a positive r is, the (smaller/larger) ______ will be the angle between the two regression lines.

__________ 41. When the correlation is zero, the regression line of $\tilde{Y}$ makes an angle of ______ degrees with the horizontal axis.

__________ 42. The quantity $\sqrt{1 - r^2}$ is called the ______. It gives us a measure of the extent to which ______ are ______. It is symbolized by the letter ______.

____

____

____

__________ 43. The value of E in the expression $E = 1 - \sqrt{1 - r^2}$ is called the coefficient of ______.

__________ 44. Table 9.9 in the text was originally published by the two men for whom it was named. Their last names were ______ and ______.

____

__________ 45. The expression r^2 indicates the proportion of the variance of Y which (can/cannot) ______ be predicted from X. It is sometimes called the coefficient of ______.

____

SAME OR DIFFERENT TYPE ITEMS

In the next four items (46 through 49), indicate whether the following pairs of symbolic terms are equivalent by responding "Yes" or "No."

______ 46. r_{xy} r_{yx}

______ 47. b_{yx} b_{xy}

______ 48. $\tilde{z}_y$ $r_{yx}z_y$

______ 49. $\tilde{z}_x$ $r_{xy}z_y$

PROBLEMS

50. If $s_x = 11.11$; $s_y = 2.22$; $\bar{X} = 33.3$; $\bar{Y} = 44.4$; $N = 55$; $r_{xy} = .66$; what is the probable X score for Percival, whose standard score in Y is -2.5?

51. If $\bar{Y} = 1.20$; $Y = 1.60$; $s_y = .20$; $\bar{X} = 20.00$; $s_x = 15.00$; and $r = -.80$; what is the predicted value of the standard score in X?

52. A girl in Oskaloosa has a score of 310 on one of two forms of the College Board Scholastic Aptitude Test. Assume that both forms have means of 500 and standard deviations of 100, and that the correlation between them is .90. Ignoring any possible practice effect, what is the most probable score that she will obtain when she is retested on the other form?

53. If the standard deviation of X is .10, the standard deviation of Y is .20, and r is $-.30$, what is the numerical value of the regression coefficient for use in predicting Y from X?

54. If $\bar{X} = 40$; $X = 50$; $x = 10$; $s_x = 5$; $\bar{Y} = 2.0$; $s_y = .4$; and $r = -.60$; what is the predicted value of y?

55. State as precisely as you can the coefficient of correlation that would arise if the standard deviation of X is double that of Y, if the mean of the X scores is three times that of the Y scores, and if the regression lines cross each other at right angles.

56. Assume the following data for 120 college students:

$x' = 0$ for the interval of 130–139 pounds
$y' = 0$ for the interval of 66–67 inches

$\Sigma x' = 22$ $\quad \Sigma y' = 2$
$\Sigma x'^2 = 294$ $\quad \Sigma y'^2 = 206$
$\Sigma x'y' = 146$

Write an equation for predicting a student's height by multiplying his weight in pounds by one constant and adding (or subtracting) another constant to (or from) this product. Then use this equation to predict the height of four students who weigh 120; 140; 160; and 180 pounds.

57. Assume the following data for 42 competitors on two civil service tests:

$x' = 0$ for X scores of 30–34
$y' = 0$ for Y scores of 55–59
$\Sigma x' = 111 \quad \Sigma y' = 99$
$\Sigma x'^2 = 355 \quad \Sigma y'^2 = 315$
$\Sigma x'y' = 322$

Compute r, the standard deviation of X, and the corresponding standard error of estimate.

58. a. Using Data Set A, Group I, compute the raw score regression equation for mathematics score used to predict final examination score.

b. Compute the standard error of estimate for the regression equation in part a.

59. Alvin's score in Arithmetic Computation is 81. His class of 50 has a mean of 103 and a standard deviation of 11 on this test, which correlates .8660 with a 135-item test of Arithmetic Reasoning. The class mean and standard deviation on the Arithmetic Reasoning test are 78 and 12, respectively. What is the most likely score for Alvin on Arithmetic Reasoning? The chances are 99 in 100 that his score will lie between what limits?

CHAPTER 10 MORE MEASURES OF CORRELATION

TRUE–FALSE

_______ 1. Sometimes it is not possible to compute the Pearson product moment correlation coefficient.

_______ 2. There are some occasions when it would be possible to compute the Pearson product moment correlation coefficient but it would be inappropriate to do so.

_______ 3. Triserial r is applicable when there are three continuous variables to be correlated.

_______ 4. Point biserial r is used with one variable which is continuous and another variable which is an artificial dichotomy.

_______ 5. Point biserial r actually is a Pearson r.

_______ 6. Tetrachoric r assumes underlying normality of both dichotomized variables.

_______ 7. Phi is an estimate of the Pearson r between two dichotomous variables.

_______ 8. Biserial r and tetrachoric r are both approximations to Pearson r.

_______ 9. Tetrachoric r gives an estimate of what the Pearson r would be if both variables were continuous, normally distributed, and expressed in many instead of only two classes.

_______ 10. The standard error of ρ is greater than that of an r of the same size based on the same N.

COMPLETION

____________________ 11. Records appropriate for analysis by biserial *r* consist of two variables, one continuous and the other a ______ .

____________________ 12. Biserial *r* is a more accurate estimate of Pearson *r* if the dichotomized variable is cut near the ______ .

____________________ 13. The standard error of biserial *r* is considerably (smaller/larger) ______ than the standard error of Pearson *r*.

____________________ 14. When one variable is divided into more than two parts (or categories), we can use an extension of biserial correlation which is called ______ .

____________________ 15. The data for tetrachoric *r* and the phi coefficient are contained in a table with ______ rows, ______ columns and ______ cells.

____

____

____________________ 16. A short-cut method for estimating tetrachoric *r* is provided by ______ .

____________________ 17. The two correlation coefficients which tell us what the Pearson *r actually is* between two variables are ______ and ______ .

____

____________________ 18. The two correlation coefficients which tell us what the Pearson *r would be* if the dichotomies were continuous and normally distributed are ______ and ______ .

____

____________________ 19. When both variables are transformed to ranks, it is possible to compute a correlation coefficient called the ______ .

PROBLEMS

20. In a group of 76 students in a degree nursing school program (the same students upon whom Data Set B is based) 30 of the students did not complete the second year. We shall call these the "Fail" group. Following are raw score data on the "Pass" group, "Fail" group and "Total" group on several pre-entrance tests. Compute biserial r by using one of the three Formulas, (10.1), (10.2), or (10.3), substituting raw scores X for the x' everywhere in the formula. Then check your answers by using one of the remaining two formulas.

	Pass Group		Fail Group		Total Group			
Test	ΣX_p	N_p	ΣX_q	N_q	ΣX	N	ΣX^2	
a. Aptitude	4,528	46	2,850	30	7,378	76	745,488	__________
b. Mathematics	602	46	315	30	917	76	14,183	__________
c. English	2,574	46	1,604	30	4,178	76	233,798	__________

d. Test each biserial r for significance.

21. A teacher wished to determine if the people who passed one particular item on an achievement test tended, in general, also to pass another particular item. Test the hypothesis that responses between the two items are due merely to chance factors by computing ϕ for the data below.

		Item #2		
		Fail	*Pass*	*Total*
Item #1	*Pass*	20	60	80
	Fail	25	15	40
	Total	45	75	120

____

22. A survey of hair color is made among two different national groups, which we shall call Group A and Group B. The percentage of blonds was 60% in Group A and 20% in Group B. A total sample of 200 cases was surveyed, split equally between the groups.

    a. What is the value of the ϕ coefficient?

    b. Test the hypothesis that the true ϕ in the population is equal to zero.

23. Assume that 40 of a company's 500 salesmen are Irish and that 25 of them have attended business college, whereas only 40% of the entire group of salesmen have done so. Substitute numbers to indicate how someone else could solve for some suitable measure of relationship. (Do not bother to compute the answer.)

	Other	Irish
Business college		
No business college		

24. If $N = 52$ and $\Sigma D^2 = 34567$, what is the value of the rank difference correlation coefficient?

25. Using Data Set G, compute the Spearman rank difference correlation coefficient between order of handing in a test paper and the test score. (Note that when two or more scores are tied, you are to assign the median of the ranks the scores would have had if they were not tied.)

CHAPTER 11 CHI SQUARE

TRUE–FALSE

_______ 1. Chi square has not been misused to any great extent.

_______ 2. The rule requiring observations to be independent of each other in using chi square implies that observations do not influence each other.

_______ 3. To fulfill the rule of equalizing the sums of the observed and the theoretical frequencies, it is permissible to do some adjusting of the observed frequencies.

_______ 4. It is desirable to set up the categories for a contingency table after one has looked at the data.

_______ 5. Chi square is suitable primarily for data in the form of frequencies.

_______ 6. If we should eliminate any row and any column from a contingency table, the number of cells left would be equal to the degrees of freedom.

_______ 7. In using chi square for a contingency table, the number of degrees of freedom is always less than the number of rows times the number of columns.

_______ 8. The absolute value of the discrepancy between expected frequency and observed frequency is the same for all four cells in a 2 X 2 contingency table.

_______ 9. If we apply Yates' correction for continuity, we may ignore the rule for minimum size of expected frequency.

_______ 10. As an alternative to chi square, we may compute the probability directly for a contingency table.

_______ 11. Chi square can be useful in assessing the adequacy of a curvefitting procedure.

_______ 12. For a contingency table to be suitable for chi square, the categories of at least one of the variables must be ordered.

COMPLETION

_______________ 13. In using chi square we compare two kinds of frequencies. They are the observed frequencies and the _______ frequencies.

_______________ 14. The authors' rule for small expected frequencies is that none should ever be smaller than _______ .

_______________ 15. If we tossed a coin 100 times and it came up with heads 40 times and if we used only the observed frequency of 40, we would be guilty of the error of _______ .

_______________ 16. If we compute the probability directly for a 2 X 2 contingency table, we would say that we have applied the _______ test of significance.

_______________ 17. If we tested the hypothesis of independence with chi square and obtained a chi square with a probability value greater than .99, we would conclude that the agreement between observed and expected frequencies was _______ .

PROBLEMS

18. Six of 61 graduate students in a college course in Psychology were named William, but only 3 of the 7 in undergraduate Psychology were so named. Why is it inappropriate to use chi square to determine whether this difference is significant? _______________

19. Groups of 98 men and 76 women were asked how many nights they preferred to dance on a 3-day weekend (Friday noon to Monday noon). If all possible answers are given, what is the number of degrees of freedom to be used in interpreting chi square? _______________

20. Groups of 317 Irishmen, 221 Mexicans, 107 Italians, 83 Armenians, and 82 Englishmen are questioned regarding their individual meat preferences (lamb, beef, pork, or rabbit). If chi square is computed, how many degrees of freedom are used in testing its significance? _______________

21. In a contingency table, row 2 contains twice as many cases as row 1 and column 3 contains three times as many cases as column 1. With chance alone operating, we would expect 6 cases in the cell at the intersection of row 1 and column 1. Then, how many would be expected by chance in the cell at the intersection of row 2 and column 3? __________

22. A coin was tossed and 96 students each guessed how it would come up, 68 guessing heads. What hypothesis is being tested? Test the hypothesis. What conclusions do you reach? __________

23. Twelve of the 61 graduate students in Psychology during one particular semester at a University were women, while 19 of the 71 undergraduate psychology students were women. Is this difference significant? __________

24. In a group of 300 students who took two test items, 180 passed item 1 and 90 passed item 2. There were 40 who passed both items. Test the hypothesis that performance on the two items is not related. __________

25. One hundred students were classified *independently* according to two nonquantiative variables as is indicated in the following contingency table:

Variable A	Variable B: *1*	*2*	*Sum*
1	15	25	40
2	45	15	60
Sum	60	40	100

a. Compute chi square for this table. __________

b. Determine the significance level of this chi square with appropriate *df*. __________

c. What conclusion can be reached regarding variables A and B? __________

d. Determine ϕ for this contingency table. __________

26. The following contingency (frequency) table was obtained by observing the hair and eye color of 164 college freshmen (males).

		Hair Color			
		Brown	*Black*	*Blond*	*Sum*
Eye Color	*Blue*	28	______	18	______
	Brown	41	26	______	73
	Hazel	______	______	______	32
	Sum	85	______	36	164

a. Complete the table by filling in the missing frequencies and sums.

b. Referring to the completed table in part a, how many degrees of freedom does it have? ______

c. Referring to the table in part a, what, if any, relationship is there between hair and eye color in this sample of observations? (No computations are needed to answer this question.) ______

d. Referring to the table in part a, what proportion of the freshmen possessed both blond hair and blue eyes? ______

27. A sample of 86 colleges and universities drawn from the population of teacher-training institutions in the United States was classified into four geographical regions as follows:

1. Northeast,
2. Southeast.
3. Midwest,
4. West.

The number of institutions in the sample in each region is shown below along with the expected frequencies (as calculated from knowledge of the population):

Region	Observed Frequency	Expected Frequency
1	23	22.0
2	22	22.3
3	32	28.2
4	9	13.5
Total	86	86.0

a. What is the *df*? ______

b. What is the value of χ^2? ______

c. What is the probability of obtaining a χ^2 as large by chance? ______

28. For each of the following 2 X 2 contingency tables, compute χ^2 by Formula (11.1). Then recompute χ^2 by Formula (11.9). Use this as a check; you should get the same answer. The letters in Formula (11.9) have the same meaning as they did in Chapter 10 on page 347 as shown in the following table:

B	A	A + B
D	C	C + D
B + D	A + C	N

			χ^2				χ^2
a.	100	100	________	e.	256	154	________
	100	100			183	243	
b.	80	80	________	f.	112	70	________
	20	20			75	110	
c.	100	60	________	g.	50	110	________
	60	100			20	20	
d.	90	30	________	h.	80	100	________
	30	10			40	80	

29. A sample of 585 graduates of a College of Education were classified on two variables, (1) Rank by major teaching field department and (2) Combinations of two attributes, marked extracurricular activities and honors at commencement. The observed frequencies are shown in the contingency table below:

	Combinations of Honors and Extracurricular			
Rank	*Neither*	*One*	*Both*	Total
High	51	80	39	170
Medium	112	100	23	235
Low	85	86	9	180
Total	248	266	71	585

a. What is the *df*? __________

b. What is the value of χ^2? __________

c. What is the probability of obtaining a χ^2 as large by chance? __________

30. For the following 2 X 2 contingency table use the Fisher exact test and report the results.

		Taught during 5th year after graduation	
		Yes	*No*
Extracurricular Activities	*Yes*	0	3
	No	11	6

____

CHAPTER 12 NONPARAMETRIC STATISTICS OTHER THAN CHI SQUARE

MATCHING

Match the appropriate classification of "Parametric" or "Nonparametric" with each statistical technique by printing either a "P" or an "N" in the blank.

(P) Parametric
(N) Nonparametric

______ 1. Chi square

______ 2. Exact probability test for 2 X 2 contingency table

______ 3. *F* test

______ 4. Mann–Whitney *U* test

______ 5. Median test

______ 6. Pearson product moment correlation coefficient

______ 7. Runs test

______ 8. Sign test

______ 9. Spearman rank difference correlation coefficient

______ 10. *t* test

COMPLETION

________________________ 11. Two other names which have been given to nonparametric methods are ______ methods and ______ statistics.

____

________________________ 12. The most important characteristic of nonparametric statistics is that they require fewer assumptions about the ______ from which the sample was drawn.

________________________ 13. Usually parametric methods are (less/more) ______ difficult to apply.

________________________ 14. When both parametric and nonparametric methods are available ______ methods are to be preferred.

________________________ 15. The null hypothesis tested in the sign test is that the percentage of plus signs is ______ per cent.

________________________ 16. Your authors recommend a (one-tailed/two-tailed) ______ test for the runs test.

________________________ 17. The runs test is (less/more) ______ efficient in testing differences between medians than the median test and Mann–Whitney U test.

PROBLEMS

18. A test of essential mathematics was given as a pretest and as a posttest in a statistics laboratory course of 15 students. The paired scores for each student are given below.

Pre	Post	Pre	Post	Pre	Post
68	71	57	62	51	53
68	66	57	64	50	59
67	68	56	63	49	60
65	70	54	58	27	49
62	60	51	57	21	30

Apply the sign test and comment on how much evidence there is for a gain in score on the test. ________________

19. An essential mathematics test was given to morning and afternoon sections of a statistics laboratory course. The scores in the two sections are shown below.

Morning	71	69	68	67	57	57	56	54	49	41	38
Afternoon	64	62	61	58	51	51	51	51	27	21	

a. Apply the runs test to detect any difference between the two sections on the mathematics test. ______________

b. Apply the median test to the same data. ______________

c. Apply the Mann–Whitney U test to the same data. ______________

d. Do the three tests of significance agree? Explain. ______________

CHAPTER 13 SIMPLE ANALYSIS OF VARIANCE

TRUE–FALSE

_______ 1. The F table is set up to test the significance of the sum of two variances.

_______ 2. Each mean square term is obtained by dividing the sum of squares by the degrees of freedom.

_______ 3. The mean square for among groups plus the mean square for within groups equals the total mean square.

_______ 4. In simple analysis of variance, all subgroups must have the same number of cases.

_______ 5. When $F < 1.00$ we can dispense with the F table.

_______ 6. When F is not significant, it is not appropriate to follow with any t tests.

_______ 7. The F test can be regarded as an extension of the t test.

COMPLETION

____
_______________ 8. To determine if two sample means differ significantly from each other, we would use the _______ test, while to determine if three or more sample means differ significantly we would use the _______ .

_______________ 9. The expression $ns_{\bar{X}}^2$ is called the _______ variance.

______

______

______

10. Analysis of variance is based upon the principle of obtaining two ______ of the ______ variance and determining if they differ enough to rule out the operation of ______ .

______

______

11. The formula for F for simple analysis of variance is the ______ variance divided by the ______ .

______

______

12. The relation WITHIN + AMONG = TOTAL is true both for ______ and for ______ .

13. The four assumptions underlying simple analysis of variance are

i. ______

ii. ______

iii. ______

iv. ______

______

______

______

14. The deviation of any raw score from the overall mean may be divided into two parts, the deviation of the raw score from ______ and the deviation of ______ from ______ .

______

15. When there are only two groups, t^2 = ______ .

______

16. The F test was named after ______ .

PROBLEMS

17. Often we have two or more groups of persons on whom we have made the same measurements and we wish to combine all persons into a composite group. In order to do this we need to meet the assumption that the groups (as samples) came from a common population. Test the hypothesis that the sample means came from one population by running a simple analysis of variance for the three classes, #1, #2, and #3 on each of the following variables in Data Set B.

    a. Aptitude ____________________

    b. Mathematics ____________________

    c. First Year Grade Average ____________________

18. The miles-per-gallon ratio was calculated for a car several times under each of four different driving conditions. The data are given below.

Driving Conditions	MPG Values					
Urban Driving Largely	21,	18,	17,	16,	16,	14
Combination Urban & Rural	24,	20,	20,	18,	17,	17
Rural Driving (High Speed)	23,	22,	22,	19,	19,	19
Rural Driving (Moderate Speed)	24,	24,	23,	22,	20	

By a simple analysis of variance, test the hypothesis that the four subsamples were drawn from a common population.

CHAPTER 14 STANDARD ERRORS OF DIFFERENCES

TRUE–FALSE

______ 1. If the correlation between initial and final tests is positive, the standard error of the differences will be smaller when the correct formula is used than when the incorrect formula omitting the correlational term is used.

______ 2. With the techniques in this chapter we can test the difference between two means and prove that the true difference is zero.

______ 3. Conclusions about physical measurements of girls in a beauty contest should be generalizable, roughly, to women in general of comparable age.

______ 4. When paired measurements are made on the same persons, the difference between the means is not the same as the mean of the pair differences.

______ 5. The correlation between standard deviations of two variables is approximately equal to the correlation between the original scores.

______ 6. The standard error of the difference between two standard deviations is usually smaller than the corresponding standard error of the difference between two means.

______ 7. The highest significance level ordinarily used by statisticians is .001.

______ 8. The alternative method, using pair differences, is appropriate to use for finding the standard error of differences between means when the correlation is unknown.

______ 9. The correlation term should be included in formula (14.1) for pairs of measurements on a set of identical twins.

COMPLETION

____________________ 10. When one tests the hypothesis that a sample statistic does not differ from some hypothesized parameter, such a parameter is always a (fixed point/figure subject to variation) ______ .

____________________ 11. When several samples of some fixed size are drawn successively, with replacement, from a population, the means of the samples tend to be (less/more) ______ variable than the original score values.

____________________ 12. If samples of the ages of brides and grooms are drawn from a population and the means for brides are paired with the means for grooms, the correlation between these means would be (less than/equal to/more than) ______ the correlation between the original ages.

____________________ 13. In Table 14.2 each pair difference is treated as though it were a ______ .

____________________ 14. When both methods for computing the standard error of the difference between means in Chapter 14 are applied
____________________ to the same data and there is a difference in the results, the reason would be that the disagreement is due to ______ . In order to eliminate such disagreement, we would ______ .

____________________ 15. The correlation term in formula (14.1) goes to zero when the observations in the two groups are ______ .

PROBLEMS

16. A group of 17 Irishmen take a test on the spelling of English words on St. Patrick's Day and, as might be expected, make low scores. Exactly 17 days later they take a comparable test and make a higher average score. What formula would be most appropriate in evaluating the results?

17. What conclusion can be drawn from the statements that a difference is 3.66 and the standard error of that difference is 6.63?

18. All 22 pupils in grade 2 of Public School 222 are randomly assigned to either a phonics or a nonphonics method of teaching reading. In May, standardized tests reveal a gain of 9.1 months for the phonics group and 8.4 months for the nonphonics group. The value of t is 1.72. What do you conclude regarding the two methods?

19. If the standard error of one mean is .80 and the standard error of another (uncorrelated) mean is .60, what is the standard error of the difference between the means?

20. Assume the following hypothetical data on movie attendance based on random samplings in the two high schools.

	Lock Haven	State College
Number of students	37.	50.
Average weekly attendance	3.41	2.83
Third quartile	4.50	3.62
First quartile	0	1.00
Standard deviation	1.86	1.96
Are movies shown on Sunday?	Yes	No
Correlation between attendance on successive weeks	.50	.60

Is attendance significantly lower in the borough that bans Sunday movies than in the borough that permits people to be their own judges of how they spend their Sundays?

21. When an enthusiastic, variety-loving 9-year-old boy asks for still another job to do, an intolerant mercenary adult replied, "No; it generally takes longer to show you how to do something than you spend doing it afterwards." The boy disagrees and feels persecuted. Who is right and who is wrong? (Show basis for your answer.) Here is what happened:

Operation	Showing	Doing
Pushing hand lawn mower	11 min.	6 min.
Operating weeding tool	6 min.	1 min.
Putting pine needles on driveway	3 min.	1 min.
Removing sticks and pine cones	2 min.	0 min.
Cutting dead bamboo shoots	8 min.	15 min.
Pulling weeds	11 min.	2 min.
Pushing wheelbarrow	3 min.	1 min.
Pruning bushes	6 min.	1 min.
Hoeing weeds around trees	6 min.	2 min.

22. Sophomores in 4 high schools were given an intelligence test, shown one of 4 versions of a film on snakes, and then tested on their knowledge of snakes. It was possible to select 123 sets of 4 girls, each set consisting of one girl from each school and all 4 having identical intelligence test scores. Write the formulas that you would use to determine whether the second film version was better or worse than the fourth version. (Label and define terms so that it is possible to tell exactly what you wish to do.)

23. Sixteen subjects took two psychological tests and obtained the following scores:

Subjects	Test X_1	Test X_2	$X_1 - X_2$
1	18	4	14
2	13	13	0
3	21	11	10
4	14	11	3
5	25	15	10
6	14	15	-1
7	7	11	-4
8	20	12	8
9	18	7	11
10	9	3	6
11	15	11	4
12	25	11	14
13	14	7	7
14	6	13	-7
15	12	10	2
16	9	10	-1
Sum of X	240	164	76
Sum of X^2	4112	1860	958
(Sum of X)^2	57600	26896	5776

a. Compute the t ratio testing the difference between X_1 and X_2. ____________

b. Determine the significance of t with appropriate df. ____________

c. What conclusion can be reached regarding the difference? ____________

SETS OF DATA

DATA SET A

Mathematics Pretest, Mathematics Opinion Scale, and Final Examination Score in Two Statistics Sections

Group I			Group II		
Math Pretest	*Math Opinion*	*Final Exam*	*Math Pretest*	*Math Opinion*	*Final Exam*
7	52	56	11	40	59
12	71	78	15	23	73
13	49	66	10	33	60
12	27	73	16	73	66
14	39	63	15	47	68
16	49	76	10	35	59
11	24	65	18	31	72
15	55	74	14	39	63
12	42	69	14	49	71
16	66	78	4	30	67
17	55	78	12	67	63
13	54	75	6	23	34
9	63	69	9	23	49
9	10	70	15	42	66
18	63	79	4	15	56
8	48	73	10	57	71
18	59	75	6	29	53
17	26	78	10	24	61
16	62	67	15	54	70
6	18	70	15	42	60
18	45	79	15	38	54
18	40	80	11	64	63
			3	1	49

DATA SET B

Pre-Entrance Tests and First Year Grade Average for Class # 1 of Nursing Students in a University Program

Case No.	Pretests: *Aptitude*	*Arithmetic*	*Mathematics*	*English*	First Year Grade Average
01	115	39	19	57	24
02	91	32	16	48	27
03	122	41	12	67	27
04	115	39	14	66	19
05	74	32	12	57	9
06	131	36	16	60	22
07	66	24	4	52	13
08	127	30	13	54	22
09	103	20	8	52	10
10	94	23	4	52	9
11	98	39	16	53	20
12	96	33	16	52	11
13	87	37	10	49	10
14	85	24	2	46	8
15	87	15	13	53	11
16	79	24	9	51	14
17	102	37	17	53	12
18	79	16	2	49	7
19	106	32	14	51	18
20	78	27	3	52	12
21	96	22	7	55	16
22	126	15	10	61	12

Pre-Entrance Tests and First Year Grade Average for Class #2 of Nursing Students in a University Program

	Pretests				First Year Grade Average
Case No.	*Aptitude*	*Arithmetic*	*Mathematics*	*English*	
23	118	39	32	70	30
29	101	37	14	59	28
30	132	24	13	65	24
31	69	36	7	41	12
32	55	25	10	37	17
33	98	26	10	52	32
34	90	19	5	68	17
35	118	38	11	65	20
36	86	44	23	57	16
37	75	17	5	44	20
38	123	27	6	61	18
39	71	26	0	46	12
40	91	23	8	49	10
41	89	32	13	62	13
42	98	23	8	51	12
43	91	41	23	54	12
44	108	23	14	67	30
45	90	35	12	56	10
46	77	32	18	53	26
47	80	22	9	49	19
48	127	32	19	62	25
49	114	31	13	56	12
50	83	11	2	56	29
51	98	12	1	56	17
52	48	28	12	49	8
53	109	33	14	58	13
54	109	21	4	55	14
55	98	39	24	57	30
56	128	33	19	66	27
57	134	30	22	73	38
58	111	47	21	64	24
59	70	6	7	47	13
60	92	24	7	58	23
61	105	43	24	56	15

Pre-Entrance Tests and First Year Grade Average for Class #3 of Nursing Students in a University Program

	Pretests				First Year Grade Average
Case No.	*Aptitude*	*Arithmetic*	*Mathematics*	*English*	
67	101	40	7	52	12
68	82	23	13	49	11
69	82	31	16	42	10
70	95	26	8	53	7
71	90	23	17	50	3
72	111	28	17	57	27
73	63	19	5	49	20
74	98	31	11	52	12
75	87	39	12	51	3
76	119	27	16	63	10
78	108	33	11	59	23
79	76	19	4	46	0
80	85	18	7	48	9
81	119	22	15	57	10
82	104	10	6	53	1
83	98	36	15	58	19
84	80	22	10	45	0
86	110	34	16	65	15
87	80	34	19	47	11
88	147	45	25	73	18

DATA SET C

X	f	x'	fx'	fx'^2
30–31	2	5	10	50
28–29	5	4	20	80
26–27	7	3	21	63
24–25	10	2	20	40
22–23	17	1	17	17
20–21	18	0	0	0
18–19	14	-1	-14	14
16–17	8	-2	-16	32
14–15	5	-3	-15	45
12–13	7	-4	-28	112
10–11	5	-5	-25	125
8–9	4	-6	-24	144
6–7	3	-7	-21	147
4–5	2	-8	-16	128
2–3	1	-9	- 9	81
	108		-80	1,078

DATA SET D

X	f	x'	fx'	fx'^2
69–71	1	5	5	25
66–68	2	4	8	32
63–65	4	3	12	36
60–62	7	2	14	28
57–59	13	1	13	13
54–56	16	0	0	0
51–53	11	-1	-11	11
48–50	9	-2	-18	36
45–47	8	-3	-24	72
42–44	6	-4	-24	96
39–41	7	-5	-35	175
36–38	5	-6	-30	180
33–35	4	-7	-28	196
30–32	3	-8	-24	192
27–29	2	-9	-18	162
	98		-160	1,254

DATA SET E

Test Score	f
290–299	2
280–289	
270–279	3
260–269	1
250–259	4
240–249	5
230–239	5
220–229	10
210–219	5
200–209	8
190–199	7
180–189	6
170–179	5
160–169	5
150–159	8
140–149	7
130–139	11
120–129	9
110–119	9
100–109	12
90– 99	10
80– 89	4
70– 79	2
60– 69	1
50– 59	1
	140

DATA SET F

Height (in.)	Weight (lb.)	Age (yrs.)	Sex
71	125	23	F
64	125	28	F
65	112	27	F
62	130	26	F
76	245	27	M
66	122	44	M
71	155	25	M
64	120	50	F
64	135	33	F
64	105	24	F
69	165	28	M
66	110	43	F
68	135	26	F
59	87	28	F
67	125	23	F
68	125	38	F
67	110	43	F
72	195	37	M
67	127	23	F
71	170	26	M
71	180	24	M
65	115	50	F
74	185	24	M
67	137	24	F

DATA SET G

Hand In Order	Test Score
1	43
2	35
3	42
4	27
5	35
6	39
7	35
8	36
9	39
10	39
11	35
12	35
13	40
14	33
15	44
16	43
17	36
18	41
19	46
20	47
21	43
22	32
23	47
24	33
25	41
26	37
27	36
28	26
29	32
30	36
31	30
32	43
33	38
34	46
35	37
36	31
37	37

DATA SET H

X	f
90–94	1
85–89	1
80–84	1
75–79	4
70–74	9
65–69	13
60–64	26
55–59	19
50–54	12
45–49	8
40–44	3
35–39	2
30–34	
25–29	1

ANSWERS TO MATHEMATICAL PRETEST

1. D	10. C	19. B	28. B
2. C	11. B	20. B	29. B
3. D	12. B	21. A	30. A
4. A	13. B	22. B	31. C
5. A	14. B	23. A	32. A
6. D	15. C	24. C	33. D
7. D	16. D	25. C	34. D
8. A	17. B	26. B	35. B
9. D	18. C	27. B	36. A

NOTE TO STUDENT: How many did you get right? How many did you miss or omit? If you got all 36 questions right, wonderful! If you got 32 to 35 correct, if you should ever have any difficulty in this course, it won't be due to insufficient mathematical background. If you got from 25 to 31 correct, it may be advantageous for you to review arithmetic or algebra or to get a small amount of tutoring. If you got from 20 to 24 correct, we recommend that you see your instructor during his office hours this week to get his suggestions about tutoring, reviewing, or maybe delaying the course until you can get a better mathematical background. If you got fewer than 20 correct, we recommend that you see your advisor about dropping the course, at least for this semester.

THE SELF AND ITS BRAIN
An Argument for Interactionism

By **Karl R. Popper** and **John C. Eccles**

1977. xvi, 597p. 66 illus. 10 tables. cloth

This timely work is the result of a collaboration between Popper (a philosopher) and Eccles (a neurobiologist) on the relation between the philosophy of the self and the biology of the brain. Popper examines the contradictions between dualism/interactionism and materialism/parallelism; Eccles examines the mind from the point of view of the neurology of the brain and its functional performance. The upshot is a radical and intriguing hypothesis on the interaction between mentation and the neurological events in the cerebral cortex. The volume's final section, based on the recorded conversations between the authors, reflects a dynamic exchange as they attempt to come to terms with what is perhaps the greatest riddle of man – the fact of his existence.

Contents

Part I by K. R. Popper: Materialism Transcends Itself. –The Worlds 1, 2, and 3. –Materialism Criticized. –Some Remarks on the Self. –Historical Comments on the Mind-Body Problem. –Summary.

Part II by J. C. Eccles: The Cerebral Cortex. –Conscious Perception. –Voluntary Movement. –The Language Centers of the Human Brain. –Global Lesions of the Human Cerebrum. –Circumscribed Cerebral Lesions. –The Self-Conscious Mind and the Brain. –Conscious Memory: The Cerebral Processes Concerned in Storage and Retrieval.

Part III: Dialogues Between the Two Authors.

Springer-Verlag
New York Heidelberg Berlin

GPSR Compliance

The European Union's (EU) General Product Safety Regulation (GPSR) is a set of rules that requires consumer products to be safe and our obligations to ensure this.

If you have any concerns about our products, you can contact us on ProductSafety@springernature.com

In case Publisher is established outside the EU, the EU authorized representative is:

Springer Nature Customer Service Center GmbH
Europaplatz 3
69115 Heidelberg, Germany

Batch number: 10866273

Printed by Printforce, the Netherlands